XTREME SURVIVAL

SURVIVING THE OCEAN

A&D Xtreme
BOLD HI-LO NONFICTION
An imprint of Abdo Publishing
abdobooks.com

JESSICA RUSICK

TAKE IT TO THE XTREME!

GET READY FOR AN EXTREME ADVENTURE!
THE PAGES OF THIS BOOK WILL TAKE YOU INTO
THE THRILLING WORLD OF OUTDOOR SURVIVAL.
WHEN YOU HAVE FINISHED READING THIS BOOK, TAKE THE
XTREME CHALLENGE ON PAGE 45 ABOUT WHAT YOU'VE LEARNED!

ABDOBOOKS.COM
Published by Abdo Publishing, a division of ABDO, PO Box 398166, Minneapolis, Minnesota 55439.

Printed in the United States of America, North Mankato, MN.
102023
012024

Design: Tamara JM Peterson, Mighty Media, Inc.
Editor: Katherine Chu
Cover Photographs: Andrey_Kuzmin/Shutterstock Images (ocean); vincentlecolley/Shutterstock Images (raft)
Interior Photographs: Aastels/Shutterstock Images, pp. 36-37; Andrey_Kuzmin/Shutterstock Images, p. 1 (ocean); Arena Photo UK/Adobe Stock, pp. 22-23; birdsonline/iStockphoto, pp. 28-29; Chanishka Colombage/Shutterstock Images, pp. 20-21; Charles Wollertz/Adobe Stock, pp. 4-5; Colombian Navy/AP Images, pp. 40-41; Francisco Duarte Mendes/Shutterstock Images, pp. 24-25; Iakov Kalinin/Adobe Stock, pp. 12-13; JCVStock/Adobe Stock, pp. 14-15; Julian Elliott/Adobe Stock, p. 44; Kathy Hutchins/Shutterstock Images, pp. 38-39; KieferPix/Shutterstock Images, pp. 16-17; Les Lee/Getty Images, pp. 30-31; Marco Ugarte/AP Images, pp. 10-11; MC2 RJ STRATCHKO, USN/ Wikimedia Commons, p. 43; Oleksandr Kalinichenko/Shutterstock Images, pp. 6-7; Oskari Porkka/ Shutterstock Images, p. 46; Photograph Curator/Flickr, pp. 26-27; Stefan Pircher/Shutterstock Images, pp. 32-33; Topolszczak/Shutterstock Images, pp. 34-35; TSgt Scott Reed/Wikimedia Commons, pp. 42-43; TTshutter/Shutterstock Images, pp. 18-19; Wikimedia Commons, p. 13; vincentlecolley/Shutterstock Images, p. 1 (raft); Zhuravlev Andrey/Shutterstock Images, pp. 8-9
Design Elements: Nik Merkulov/Shutterstock Images (grunge background); queezz/Shutterstock Images (brush texture)

Library of Congress Control Number: 2023939330

Publisher's Cataloging-in-Publication Data
Names: Rusick, Jessica, author.
Title: Surviving the ocean / by Jessica Rusick
Description: Minneapolis, Minnesota : Abdo Publishing, 2024 | Series: Xtreme survival | Includes online resources and index.
Identifiers: ISBN 9781098291846 (lib. bdg.) | ISBN 9781098278748 (ebook)
Subjects: LCSH: Survival--Juvenile literature. | Survival skills--Juvenile literature. | Survival at sea--Juvenile literature. | Shipwreck survival--Juvenile literature. | Outdoor life--Juvenile literature. | Wilderness survival--Juvenile literature.
Classification: DDC 613.69--dc23

TABLE OF CONTENTS

A FISHING TRIP 4

SURVIVOR SPOTLIGHT 10

PROTECTING AGAINST THE ELEMENTS. 12

FINDING WATER. 16

FINDING FOOD 24

ENCOUNTERING ANIMALS 30

KEEPING CALM 34

GETTING RESCUED. 40

SURVIVING THE OCEAN 44

XTREME CHALLENGE 45

GLOSSARY 46

ONLINE RESOURCES 47

INDEX 48

CHAPTER 1

A FISHING TRIP

November 17, 2012, began normally for José Alvarenga and Ezequiel Córdoba. The two fishermen left Costa Azul, Mexico, for a two-day trip on the Pacific Ocean. But after one day at sea, a powerful storm blew their 25-foot (7.6 m) boat off course. They quickly bailed water to keep the boat from sinking.

A fishing boat struggles in a storm. To keep their boat afloat during the storm, Alvarenga dumped anything heavy into the ocean, including the fish they had caught.

Fishing on the ocean can be dangerous. Between 2014 and 2021, an average of more than 500 marine deaths, accidents, and other incidents were reported on fishing boats each year.

The storm stopped after seven days. But it had destroyed the boat's motor and radio. This meant Alvarenga and Córdoba could not control the boat or call for help. The storm also ruined their fishing supplies, making it almost impossible to catch food. The fishermen were **stranded** in the middle of the ocean with no food or fresh water. How would they survive?

XTREME FACT

In the US, the coast guard uses computer programs to help predict where disabled boats or rafts are located. They can then send rescue craft to that location.

The ocean is a **dangerous** place to become **stranded**. There is little to no food or fresh water. Long exposure to the elements takes a harsh **toll** on the body. But many people have overcome these dangers and more, thanks to their survival skills.

Experts say that to survive longer, water and food should be conserved.

SURVIVOR
SPOTLIGHT

With no usable fishing gear, Alvarenga used his hands to catch fish, turtles, and seabirds for food. When it rained, the fishermen caught drinking water with a bucket. They promised to pass messages to each other's family if one of them did not survive.

Unfortunately, Córdoba died after two months. But Alvarenga survived. He imagined detailed fantasies to occupy himself. After 438 days, his boat drifted close to land. Alvarenga swam to shore and was rescued!

Alvarenga (*center*) with his father, Jose Ricardo (*left*), and mother, Maria Julia (*right*), while on his way to give Córdoba's message to his mother in Mexico

CHAPTER 2

PROTECTING AGAINST THE ELEMENTS

Having shelter is one of the keys to surviving on the ocean. The sun and ocean salt water can make skin crack and bleed. It's important to keep skin covered to protect it from sunburn. Rain and wind can also make nights bitterly cold. So, it's important to stay as dry and warm as possible.

Chinese sailor Poon Lim was stranded on a wooden raft after the merchant ship he was on sank in 1942. He created a shelter using canvas.

Even a simple shelter can help protect you from the elements. In 2010, teenagers Filo Filo, Etueni Nasau, and Samu Tonuia, from the New Zealand territory Tokelau, were **stranded** in a small boat for 51 days. The boys wrapped themselves in a tarp to stay warm at night and dry during a storm.

If possible, hang clothes on hot days. This will help dry the clothes out and can provide shade.

CHAPTER 3
FINDING WATER

Finding drinking water is also important to ocean survival. The human body can only survive a few days without water! Water helps keep the body's cells alive and protects organs from damage. Lack of water is called dehydration.

Dehydration can cause hallucinations, seizures, kidney failure, and ultimately death.

The ocean is made up of water. But seawater is too salty for humans to drink. Doing so causes water to leech out of the body's cells, quickly leading to dehydration. Luckily, there are ways to get fresh water at sea. The healthiest source in an emergency is rainwater.

Water can also be collected if there is fog. Use clothing to collect water droplets off the sides of the boat or raft, then wring the water into your mouth.

XTREME FACT

If you can, cover a container of seawater and trap fresh water as the seawater evaporates. This will desalinate the water and make it safe to drink!

Being on a boat without shelter in intense sunlight for an extended time can lead to a dangerous sunburn.

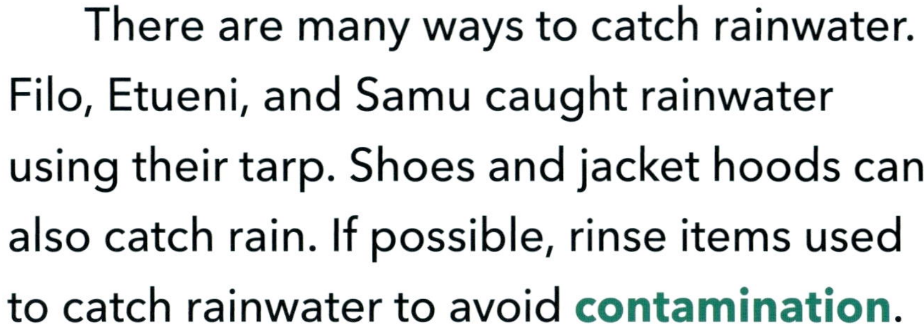

There are many ways to catch rainwater. Filo, Etueni, and Samu caught rainwater using their tarp. Shoes and jacket hoods can also catch rain. If possible, rinse items used to catch rainwater to avoid **contamination**.

XTREME FACT

Humans lose water through sweat. To avoid dehydration, it's best to stay cool and limit activity.

The seabirds that Alvarenga caught provided him and Córdoba with food and water.

When there's no rain, survivors must be resourceful. Alvarenga drank seabird blood to survive between rainstorms. The blood contained enough water to keep him hydrated. However, animal blood contains other compounds that can make people sick. So, it is best to only drink it if there are no other options.

XTREME FACT

Drinking urine is not a good way to stay hydrated. Urine contains salt, which dehydrates the body.

CHAPTER 4

FINDING FOOD

Finding food is also important to survival. In 2015, American sailor Louis Jordan was **stranded** at sea for 66 days. At first, he had difficulty catching fish. One day, Jordan was cleaning his shirt in the ocean and noticed it attracted fish! So, Jordan used his shirt as a net to catch fish. Clothes can also be used to gather plankton.

XTREME FACT

Digestion requires water. So, it is unwise to eat when severely dehydrated.

If possible, use a flashy object as a lure to catch fish. Then use the fish guts as bait to catch more.

Some survivors are able to craft fishing supplies. In the 1940s, Chinese sailor Poon Lim was **stranded** on a raft for 133 days. Lim made fishing hooks and line from a flashlight spring, nail, and rope. When he caught a small fish, he used it as bait to attract larger fish.

Lim (*right*) shares with Rear Admiral Julius Furer how he created a fishhook out of a flashlight spring.

Fish are not the only source of food on the ocean. Floating seaweed may contain crabs and shrimp. Seabirds may also land on boats or rafts to rest. Since most seabirds have very few predators, they are not scared of humans. This makes it possible to catch them by hand.

Seabirds, such as the Wilson's petrel, can be lured by bait and caught by hand, hit with an oar, or caught with a net or hook.

CHAPTER 5

ENCOUNTERING ANIMALS

In 1973, British couple Maurice and Maralyn Bailey were **stranded** on a life raft for over 117 days. During that time, barnacles grew on the bottom of the raft, attracting sea life. This made it easier for the Baileys to catch food. And watching the animals kept their minds occupied!

A whale damaged Maurice (*right*) and Maralyn (*left*) Bailey's yacht. Luckily, they were able to gather some food and supplies into their life raft before the yacht sank.

However, being surrounded by that much sea life can attract sharks. Though sharks are unlikely to attack, splashing and loud noises can **provoke** them. Stay still and quiet when they are nearby. Maintain eye contact with the shark to show **dominance**. If the shark attacks, hit it in the gills. This makes it harder for the shark to breathe.

Seals are a shark's natural food source. Sometimes sharks attack humans because they mistake them for seals.

CHAPTER 6
KEEPING CALM

Experts say that staying positive and having a strong will to live is helpful when trying to survive.

Attitude is perhaps the most important survival skill of all. Being **stranded** at sea is scary. It can cause people to panic and give up hope. Survivors must **focus** on staying calm and alive instead of giving in to sadness, anger, or fear.

In 1983, American sailor Tami Oldham Ashcraft was sailing on the Pacific Ocean with her fiancé, Richard Sharp. When a hurricane hit their boat, Sharp was swept out to sea and Ashcraft was knocked **unconscious**. She woke up after 27 hours, alone in a damaged boat.

Ashcraft and Sharp were experienced sailors. Although Sharp had a safety harness on when the storm hit, it did not save him from being thrown overboard.

Ashcraft (*second from left*) at the *Adrift* film premiere. In 1998, she wrote a memoir called *Red Sky in Mourning* about her experience, which inspired the 2018 film.

Losing Sharp **devastated** Ashcraft. But she **focused** on survival. She made a sail to guide the boat toward Hawaii. And she calculated her location using the stars and sun. Doing this math helped keep her calm and focused. After 41 days, Ashcraft was rescued near Hawaii!

SHAILENE WOODLEY SAM CLAFLIN
ADRIFT
BASED ON THE INCREDIBLE TRUE STORY
BASED ON THE BOOK BY TAMI OLDHAM ASHCRAFT WITH SUSEA MCGEARHART
SCREENPLAY BY AARON KANDELL & JORDAN KANDELL AND DAVID BRANSON SMITH
DIRECTED BY BALTASAR KORMÁKUR
INGENIOUS
www.Adrift.movie

CHAPTER 7

GETTING RESCUED

Hailing a passing ship or plane can lead to an ocean rescue. In late 2022, Dominican sailor Elvis Francois became **stranded** on the Caribbean Sea. Using a mirror, he reflected light at a passing plane and wrote "help" in large letters on his sailboat. This helped Francois get noticed and rescued!

XTREME FACT

Reflective objects, such as cell phone screens, can also be used to catch the attention of a passing vessel.

Francois (*right*) being checked by a Colombian Navy member after being rescued

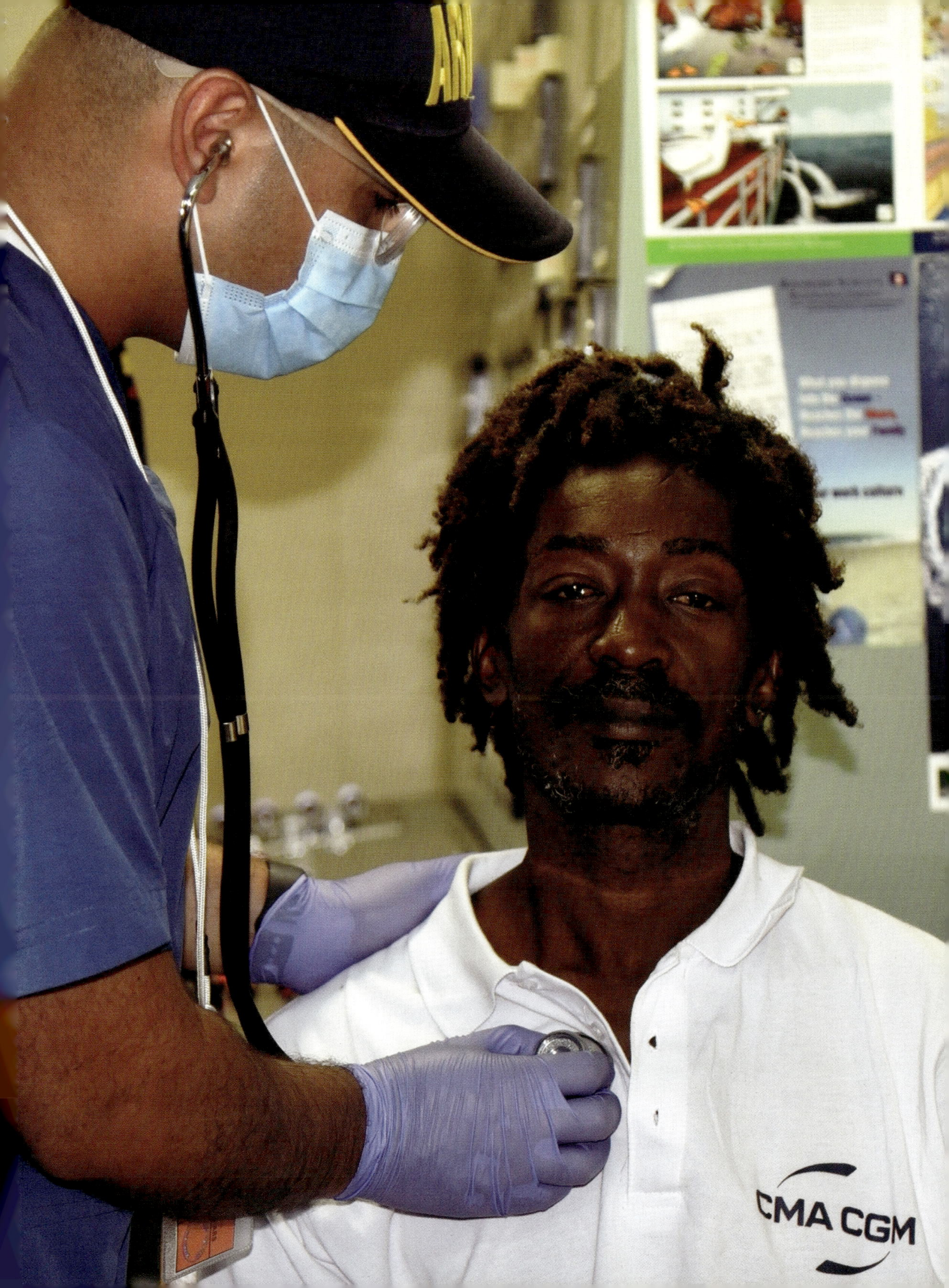
CMA CGM

Bright, reflective objects can help signal a boat or plane. But it's often hard to spot small boats. Many rescues require the luck of drifting near land. It's best to save energy in case swimming to land becomes an option. Look for signs that land is near. These signs include shorebirds and pieces of driftwood.

In 2008, a US Navy ship, the USS *Fort McHenry* (*lower left*), rescued two stranded fishermen (*upper right*) who got the Navy's attention by waving their shirts and rubbing their stomachs.

CHAPTER 8

SURVIVING THE OCEAN

Being **stranded** on the ocean means having little **access** to fresh water, food, and shelter. Storms and sharks are constant dangers. But with top-notch survival skills and a bit of luck, some people have overcome these challenges to survive. Think back on the ocean survival stories you read and skills you learned. Could you survive on the ocean?

Wear bright clothes when riding in a boat. This can help rescuers spot you if you are stranded on the ocean.

XTREME CHALLENGE

1) How did José Alvarenga get fresh water when he was stranded?

2) What did Elvis Francois use to reflect light at potential rescuers?

3) Why is it dangerous for humans to drink seawater?

4) If you were stranded on the ocean, would you rather have a tarp or fishing supplies? Why?

5) How would you keep your mind occupied if you were stranded on the ocean?

GLOSSARY

access–the opportunity to gain or use something.

attitude–the way you think or feel about something.

contamination–the process of making something unfit for use by adding something harmful or unpleasant.

dangerous–able or likely to cause hurt or harm.

desalinate–to remove the salt.

devastate–to overwhelm with grief.

digestion–the process of breaking down food into simpler substances the body can absorb.

dominance–the state of having power over another or others.

evaporate–to change from a liquid into a gas or vapor.

focus–to concentrate on or pay particular attention to.

predict—to guess something ahead of time on the basis of observation, experience, or reasoning.

provoke—to cause a person or animal to become angry or violent.

stranded—lacking the means to leave a place.

toll—a cost in life or health.

unconscious—the state of not being awake and losing awareness of one's surroundings.

ONLINE RESOURCES

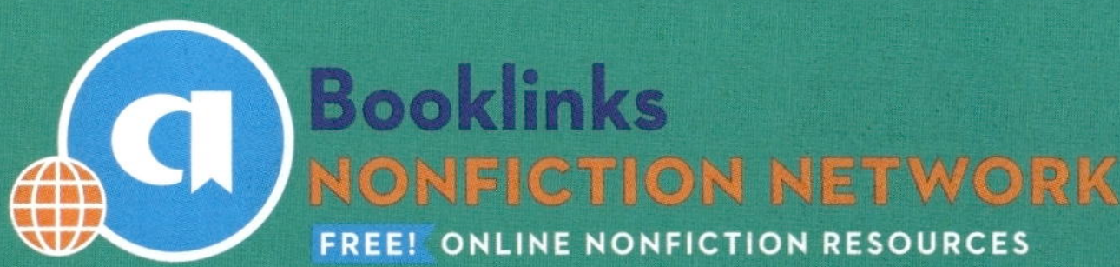

To learn more about ocean survival, please visit **abdobooklinks.com** or scan this QR code. These links are routinely monitored and updated to provide the most current information available.

INDEX

A
Alvarenga, José, 5, 7, 11, 23
Ashcraft, Tami Oldham, 36, 38
attitude, 11, 31, 35, 38

B
Bailey, Maralyn, 31
Bailey, Maurice, 31
bait, 26
blood, 23
boats, 5, 7, 11, 14, 28, 36, 38, 40, 42

C
Caribbean Sea, 40
clothes, 21, 24
Córdoba, Ezequiel, 5, 7, 11
Costa Azul, 5

D
dehydration, 17, 18, 21, 23, 24
desalination, 19
digestion, 24

E
elements, 8, 12, 14

F
Filo, Filo, 14, 21,
fishing, 5, 7, 11, 24, 26
food, 7, 8, 11, 24, 26, 28, 31, 44
Francois, Elvis, 40

H
Hawaii, 38
human body, 8, 17, 18, 21, 23
hurricanes, 36
hydration, 23

J
Jordan, Louis, 24

L
land, 11, 42
Lim, Poon, 26

M
Mexico, 5

N
Nasau, Etueni, 14, 21

P
Pacific Ocean, 5, 36
plankton, 24

R
rafts, 7, 26, 28, 31
rainwater, 11, 18, 21
reflective objects, 40, 42
rescue, 7, 11, 38, 40, 42
rescue craft, 40, 42

S
sails, 38
sea life, 11, 23, 24, 26, 28, 31, 32, 44
seabirds, 11, 23, 28
seaweed, 28
sharks, 32, 44
Sharp, Richard, 36, 38
shelter, 12, 14, 44
shorebirds, 42
signaling, 40, 42
storms, 5, 7, 14, 23, 36, 44
sunburns, 12
supplies, 7, 11, 26, 42
sweat, 21

T
tarps, 14, 21
Tonuia, Samu, 14, 21

U
urine, 23
US Coast Guard, 7

W
water, 7, 8, 11, 17, 18, 21, 24, 44